CONSTELLATIONS

Rebecca Woodbury, Ph.D., M.Ed.

Gravitas Publications Inc.

CONSTELLATIONS

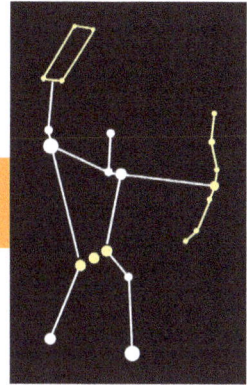

Illustrations: Janet Moneymaker

Constellations
ISBN 978-1-950415-43-4

Published by Gravitas Publications Inc.
Imprint: Real Science-4-Kids
www.gravitaspublications.com
www.realscience4kids.com

RS4K

Photo credits: Cover, Title Pg, & P.21: vchalup, AdobeStock; P.3. karandaev, AdobeStock;

When you look at the night sky,
you can see many stars.

I love looking at the stars!

A group of stars that forms a shape
in the sky is called a **constellation.**

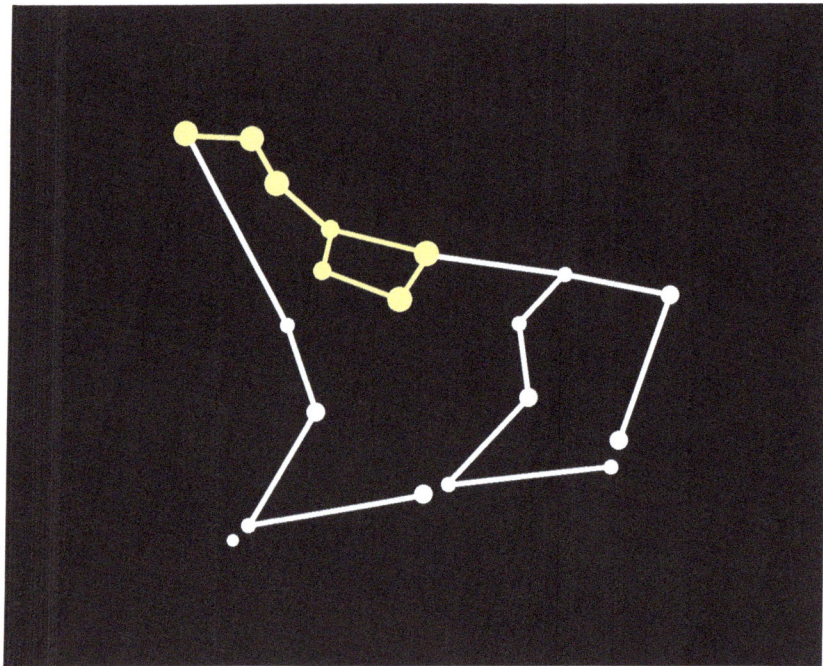

It's a bear!

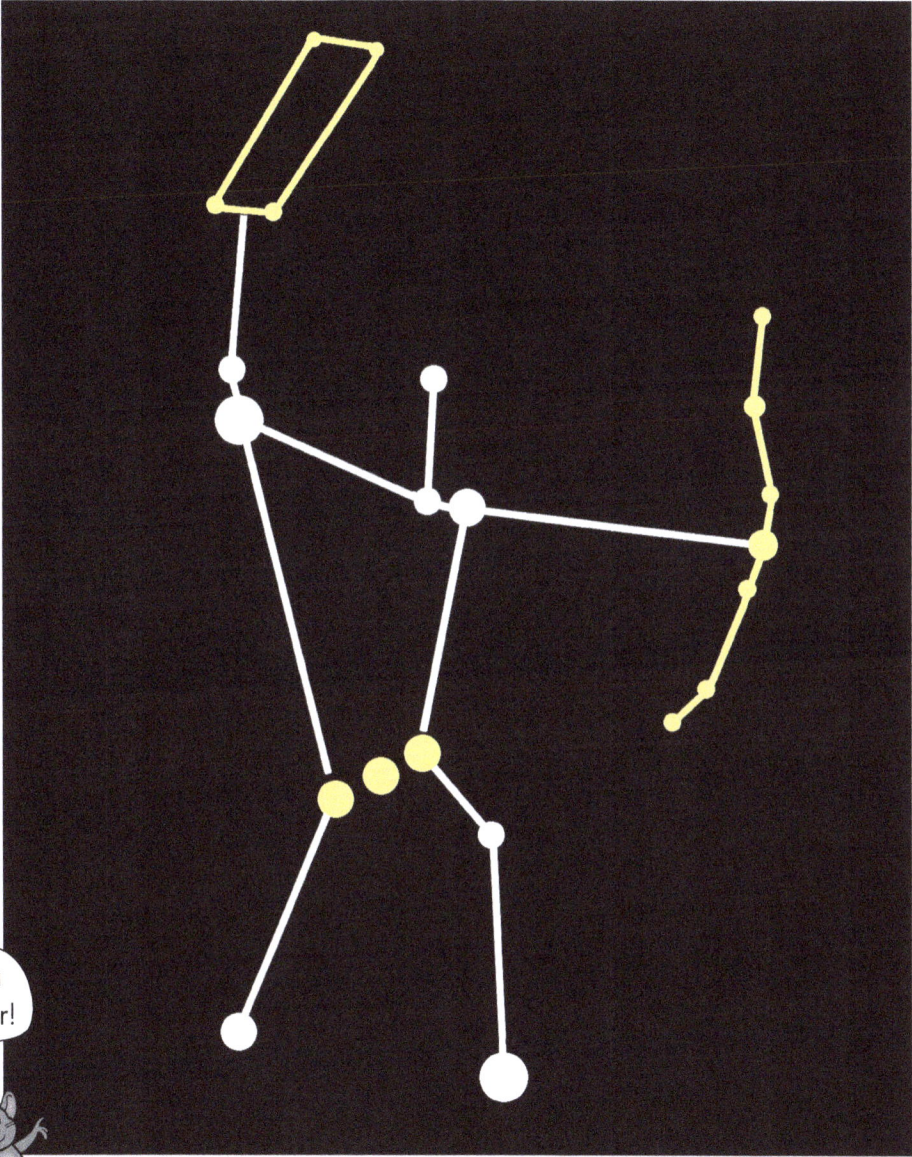

People have given names to
the different constellations.

I think it looks like cheese!

The constellations you can see are different depending on where you live.

The **Northern Hemisphere** is the northern half of Earth. The **Southern Hemisphere** is the southern half of Earth.

Which hemisphere are we in?

Hmmm. I don't know.

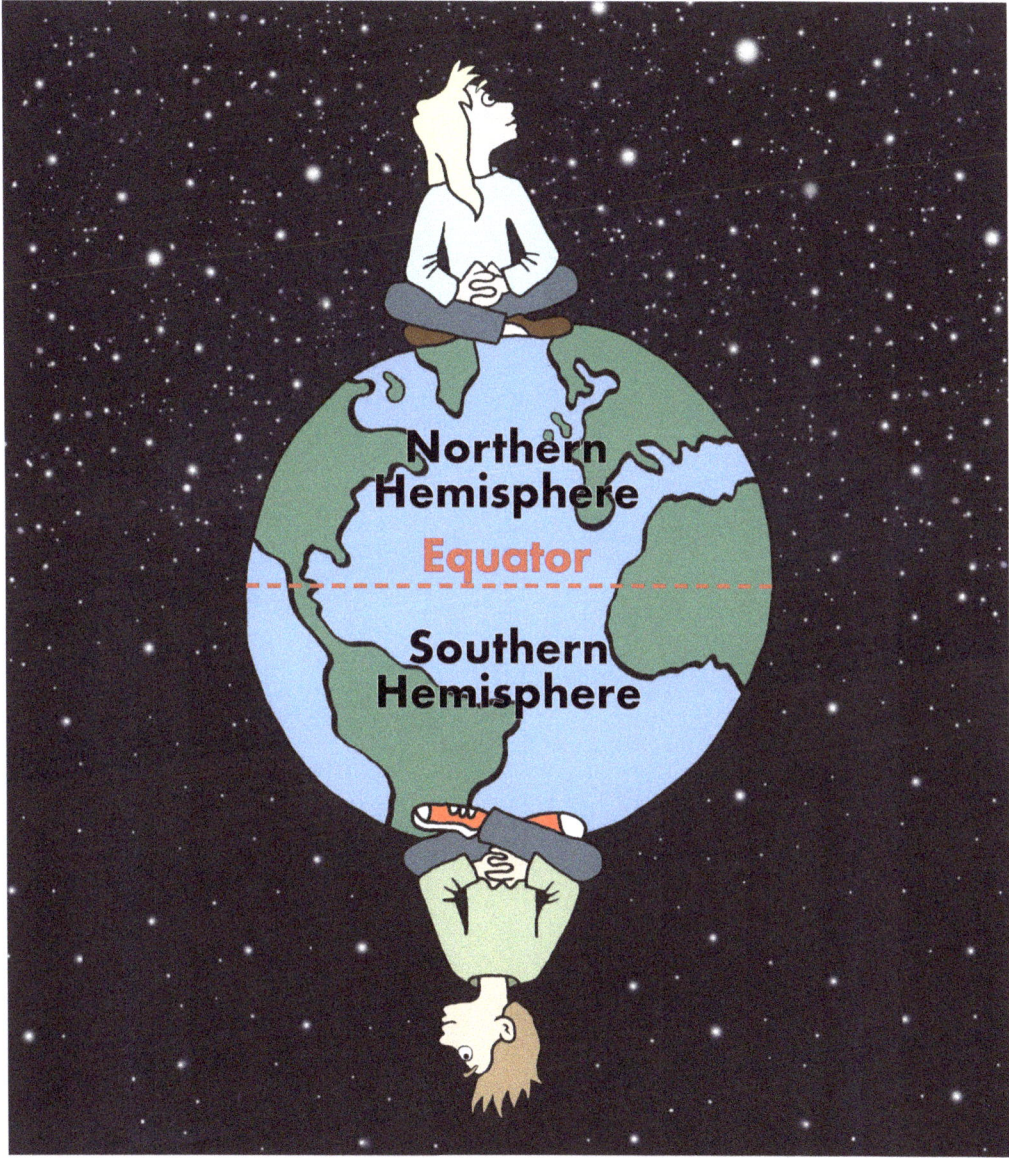

From the Northern Hemisphere
you can see the **Big Dipper**
and **Cassiopeia.**

I have seen that
big W in the sky!

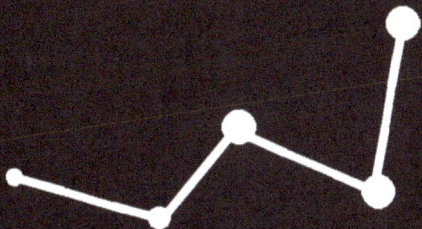

Cassiopeia

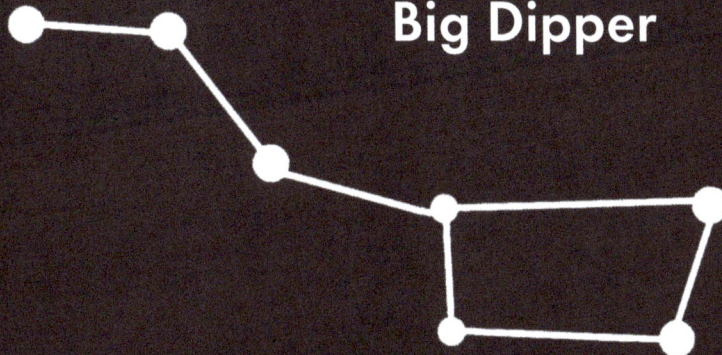

Big Dipper

From the Southern Hemisphere
you can see the **Great Square,**
Pegasus, and the **Southern Cross.**

Pegasus looks
like a horse!

The Great Square

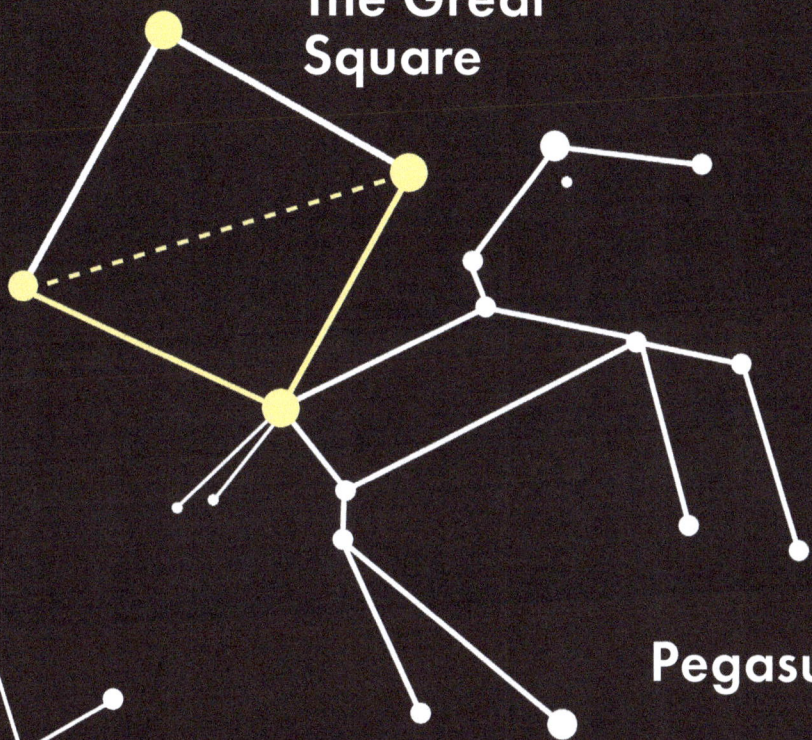

Pegasus

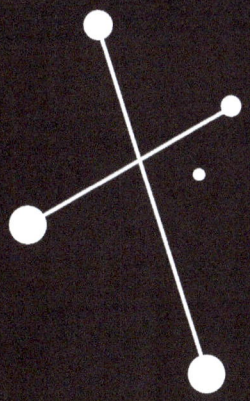

The Southern Cross

If you live at the **equator,** you can see all the constellations.

The Big Dipper points to **Polaris**.
Polaris is also called the **North Star.**

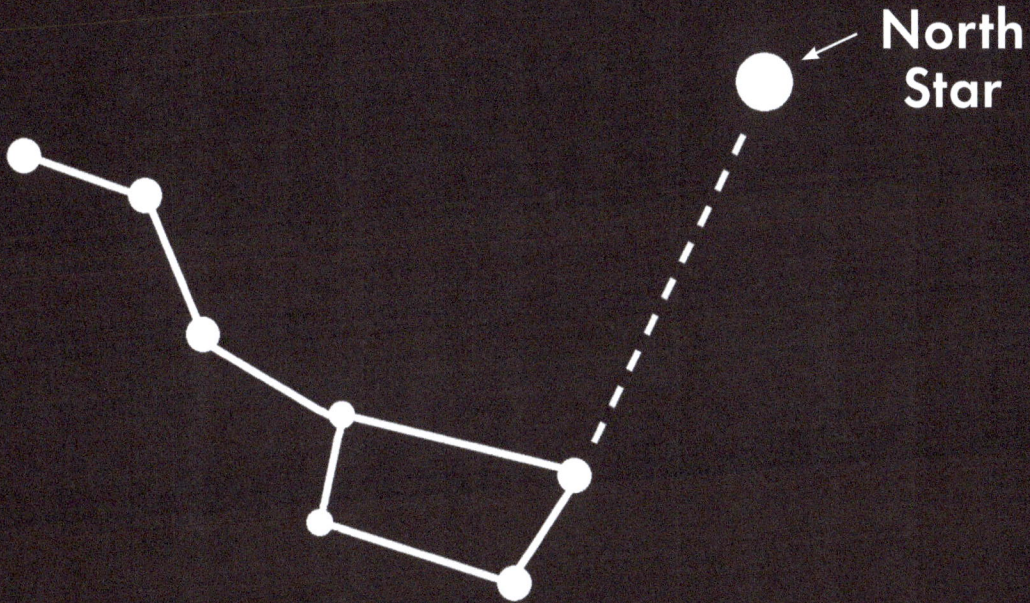

Big Dipper

North Star

Polaris is called the North Star because it is always above the northernmost part of Earth. People in the Northern Hemisphere can use the North Star to help them find their way.

People in the Southern Hemisphere can use the Southern Cross to help them find their way.

Learning to recognize the constellations is a fun way to spend time outdoors.

Look! I see the constellation called The Mouse!

I don't think there is one.

How to say science words

Big Dipper   (BIG DIH-puhr)

Cassiopeia   (kaa-see-uh-PEE-uh)

constellation   (kahn-stuh-LAY-shuhn)

equator   (ih-KWAY-tuhr)

Great Square   (GRAYT SKWEHR)

North Star   (NAWRTH STAHR)

Northern Hemisphere   (NAWR-thurn HEH-muh-sfir)

Pegasus   (PEH-guh-suhs)

Polaris   (puh-LER-uhs)

science   (SIY-uhns)

Southern Cross   (SUH-thuhrn KRAWSS)

Southern Hemisphere   (SUH-thuhrn HEH-muh-sfir)

www.ingramcontent.com/pod-product-compliance
Lightning Source LLC
Chambersburg PA
CBHW040149200326
41520CB00028B/7540